RICE HUSK ASH CEMENT

Progress in Development and Application

A report on site visits to

India, Nepal and Pakistan

by   R. G. Smith
Building Research
Establishment
Watford.

Practical
ACTION
PUBLISHING

Practical Action Publishing Ltd
27a Albert Street, Rugby, CV21 2SG, Warwickshire, UK
www.practicalactionpublishing.org

© Intermediate Technology Publications 1984

First published 1984\ Digitised 2008

ISBN 13 Paperback: 978 1 85339 051 7
ISBN Library Ebook: 978 1 78044 424 6
Book DOI: http://dx.doi.org/10.3362/9781780444246

Since 1974, Practical Action Publishing has published and disseminated books and
information in support of international development work throughout the world. Practical
Action Publishing is a trading name of Practical Action Publishing Ltd (Company Reg.
No. 1159018), the wholly owned publishing company of Practical Action. Practical Action
Publishing trades only in support of its parent charity objectives and any profits are
covenanted back to Practical Action (Charity Reg. No. 247257, Group VAT Registration
No. 880 9924 76).

PREFACE

The use of the ash resulting from burning rice
husk as a cementitious binder in conjunction with
lime has been receiving increasing attention over
recent years. It would appear to offer the
attractive prospect of providing a low-cost
alternative cement by simple, village-scale
technology using a readily available agricultural
waste material. The need for such materials is
clear as the conventional binder, Portland cement,
is commonly unavailable or expensive particularly
in rural communities where, indeed, its high-
strength, rapid-setting properties are not of
primary importance to the low-rise building or
simple civil engineering work typically undertaken
in these areas.

The basic research and development work into rice
husk ash (RHA) cement technology has been
undertaken by a variety of institutions and
individuals, principally in the Indian
sub-continent, and several attempts have been made
to commercialise and disseminate the technology.
Reports on this work are sparse, and in particular
there has been a lack of independent comparative
data on the various approaches used; these, it
will be seen from this report, contain significant
differences in terms of the final product and the
production process used. These differing
approaches were adopted for good reason and, as in
all appropriate technologies, it is unlikely that
a single process will emerge as being the unique
solution for all situations. This report is not,
therefore, intended to be a guide to the "best
buy" in RHA cement technology but is an attempt to
bring together comparative data on the various
initiatives for the benefit of those contemplating
the use of such a material in their own region.

The report must be read with all the usual
cautions; in particular the laboratory tests were
performed on 'grab' samples and should not be seen
as definitive or statistically significant.

Our thanks are due to those people and
institutions visited who gave their time and
shared their experiences freely with the compilers
of this report.

F. R. Almond, Intermediate Technology Industrial
Services, February 1984.

SUMMARY

The purpose of the four-week visit was to conduct a 'state-of-the-art' review of rice husk ash cement technology and production in the Indian sub-continent. Twelve specific locations were visited to discuss research and development. Visits to pilot and commercial production plants were included, and samples were collected for later analysis in the UK.

Some of the completed research still awaited application but some was being applied. Commercial production was often held up through non-delivery of raw material, failure of power supplies or mechanical breakdown, but some rice husk ash cement was being made, sold and used.

Research workers had found that it was necessary to limit the temperature of the burning husks to no more than $700^{\circ}C$ to obtain an amorphous ash which would be reactive with lime, although in some production processes this temperature was exceeded.

It had been found that the ash needed several hours of grinding to produce a sufficiently fine powder. The period in practice using ball mills varied between 1.5 and 5 hours.

In most cases the cement was made by mixing one part of lime with two parts of rice husk ash (by weight). Sometimes ordinary Portland cement was used in place of lime. It was commonly recommended that for building use, one part of rice husk ash cement should be mixed with 3 parts sand (by volume).

The rice husk ash cement is much cheaper than ordinary Portland cement for a given weight and in spite of richer mixes being used in practice it is still cheaper to make mortars from the rice husk ash cement. Capital investment of some £12,000 may be required for a complete plant.

Results of tests in UK laboratories are given and are compared with manufacturers' claims and with requirements in the most relevant Indian and British Standard Specifications.

The rice husk ash cements often exceeded the requirements of the Indian Standard for Lime-Pozzolana Mixture. In general they should be suitable for use in making mortars, renderings for walls, and some lower strength requirement concrete.

It is recommended that the technology be applied in other rice-growing countries of the world.

CONTENTS

KEY TO INSTITUTES

INDIA

A   Cement Research Institute of India, New Delhi

B   Indian Institute of Technology, Kanpur

C   Birla Institute of Scientific Research, Ranchi

D   Central Building Research Institute, Roorkee

NEPAL

E   Research Centre for Applied Science and Technology, Kathmandu

PAKISTAN

F   Pakistan Council for Scientific and Industrial Research, Karachi

1.   INTRODUCTION

Throughout the world eighty million tonnes of husk become available each year during the processing of rice. Some is used as packaging or fuel, but the greatest proportion is dumped as a waste by-product.

Several institutes have investigated the properties of the ash of the burnt husk, which contains a high proportion of silica, and there has been some commercial exploitation of its pozzolanic reaction with lime to form a cementitious material of potential use in the building and construction industry. Research was carried out in the United States by P. K. Mehta, and later the first meeting of the United Nations (ESCAP) National Focal Points for Regional Centre for Technology Transfer (Bangalore, India), in 1978 decided that RCTT should promote rice husk ash (RHA) cement technology. In consequence, three international RHA cement workshops have taken place; at Peshawar (Pakistan 1979), Alor Setar (Malaysia 1979) and Delhi (India 1981), where information was presented and discussed. Training centres, compilation of a Standard Specification, study of use in building and preparation of a manual on manufacture were recommended. An information centre has been established at Bangalore.

The Intermediate Technology Industrial Services (ITIS) unit of the Intermediate Technology Development Group requested a review of the current state of this art of producing a cementitious material from rice husks in the Indian sub-continent. It was already known that development work had been done there by several organisations. Various claims, reports and publications gave a confusing picture of the situation and there was no overall comparative review of the subject. ITIS recognised the potential benefits of making these materials in India and also in other rice growing areas of developing countries to increase the supply of building materials, especially where ordinary Portland cement was expensive or scarce. Such use would also effectively dispose of and utilise, an agricultural waste, improve rural economy and generate employment, with low capital costs. ITIS wished to be in a position to make recommendations so, in 1982 the author, a member of the staff of the Overseas Division, BRE, visited India, Nepal and Pakistan to gain first-hand information.

Professor D. J. Cook, Associate Professor in the
Department of Civil Engineering at the University
of New South Wales, accompanied the author for
part of the visit.

2. SITE VISITS

Discussions on research into the various
processes were held at six institutes and visits
were made to inspect the commercial enterprises
utilising the process of each institute. The map
(Figure 1) shows the location of places visited
and other centres of information. This section of
the report describes discussions and site visits
but a more detailed comparison of methods,
properties of the products and economics is made
in later sections.

Conversion of local currencies into Sterling is
based upon rates operating at the time of the
visit namely: £1 equivalent to Indian Rs. 16,
Nepalese Rs. 24, Pakistani Rs. 19.

2.1 INDIA

Research Institute A

Small experimental units for burning husk had been
built and operated and 'know how' and services had
been sold to entrepreneurs. It was claimed that a
consistently good product was best obtained by
controlling the burning temperature of husks to
less than about 600 to 650°C, in order to produce
active amorphous ash. Higher temperatures were
said to result in a crystalline less-reactive ash.

Ready-burnt ash such as that obtained from
parboiling of rice had been found unsatisfactory,
being up to 95% crystalline. Early laboratory
experiments with a round brick-built incinerator
with little ventilation (Figure 2) did not produce
good ash. A small square incinerator (Figure 3)
of open brickwork with metal mosquito netting to
retain husks and ash had the advantages of less
thermal mass, and greater access for air. A
satisfactory reactive ash was claimed to be
produced. The fairly high carbon content of the
ash produced did not prevent it from being
acceptable. Different sources of husk had shown
variations in ash residue after burning, from 20
to 28% by weight of husk, and silica content of
the ash was from 82 to 96%, although all these
could be used satisfactorily. In order to

obtain the desired high standard it was recommended that, for commercial manufacture, the lime be produced at the same location, limestone being brought in and burnt in kilns under the same careful supervision as used in burning husks, rather than purchasing ready-made lime from the existing market. Hydraulic lime, which can produce some calcium silicate itself, was found satisfactory. Too high a proportion of lime in admixture with ash was said to delay the development of strength of the product in use, and a weight ratio of 1:2 lime:ash was used.

To obtain sufficiently fast reaction rates between ash and lime it was necessary to grind the ash in a ball mill for one hour, then add lime, and continue milling for a further half hour to mix the components. Entrepreneurs were encouraged to mill the ash for 1.5 hours to be more certain of attaining the necessary fineness. They were also advised to obtain a ball mill of slightly larger capacity than that required for continuous use, to allow for breakdowns and the inevitable interruptions in electric power supply. Such a mill is available in India ready-made, its size being 1.5 x 1.5 m with a 15kW motor.

A two tonne per day output, which may be suitable for rural use adjacent to a rice mill, was reckoned to be the minimum economic size. Such a plant would be run by nine people (one supervisor, one lime kiln operator and 7 unskilled labourers). The Institute listed items of capital expenditure, of which the ball mill complete with ball charge was the most expensive at £5,000, followed by an allocation of nearly £4,000 for civil works which included boundary walls, storage and equipment accommodation, and the cost of 1000m^2 of land at £1,200. A lime kiln and husk incinerator (scaled up from the earlier experimental design) were each priced at £500. Other items brought the total of fixed capital to £12,000.

The annual costs of running the plant for 330 days per year at 90% capacity were also indicated. Raw materials were a small part of the whole: cost for rice husk was taken as charges for transportation only, so the 1950 tonnes required per annum would cost £1,180. 300 tonnes of limestone plus transport would cost £1,800. Cinder for burning the limestone would cost £1,300; electricity, chiefly for the ball mill, was calculated at £220 but this seems to be an underestimate. Depreciation of equipment would be anticipated at £750, largely due to wear of the ball mill shell, which might last only three years. Salaries of £2,250, and similar amounts for interest on loans and for packaging plus a number of smaller items would give a total annual cost of £12,000 or £19 per tonne. With an estimated selling price of £27 per tonne this would yield a return on investment of 32%.

The above selling price was much lower than the lowest price for ordinary Portland cement (OPC) of at least £40 per tonne. One third of the OPC production in India must now be sold on the open market (since February 1982), and it commands an even higher price, up to about £95 per tonne. RHA cement therefore appears very attractive economically.

Agreements had been signed with entrepreneurs in four States for plants with 4, 4, 3 and 10 tonnes per day capacities respectively. Agreements are made with a payment by entrepreneurs for plant commissioning, routine testing, examination of samples for the first month or so, and occasional testing and advice later. Laboratory tests were to ASTM, for water retention, water demand, fineness, compressive strength, slump (on flow table), and lime reactivity of ash (at 50°C). No specific standard for RHA cement existed but the products were generally assessed by the requirements of IS 4098. Specification for Lime - Pozzolana Mixture. There was little feedback of detailed information from plants at the time of the visit, except from one which was visited later, as part of the review. The larger plant which had two mills, each 1.8 x 1.8 m, was reported to be not working, due to some mechanical failure.

Six more parties were about to sign agreements and a further 400 enquiries had been received.

It was envisaged that in rural communities village people would bring their rice to a local mill, and could purchase RHA cement from the adjacent plant, taking it to their villages using the same transport as for the rice.

Experiments carried out by the Institute demonstrated the fairly successful use of RHA cement:sand mixes for renderings (1:3 by volume). Although cracks had appeared in some elevations, the renderings seemed well attached to the brickwork. Fair-faced brickwork built with RHA cement:sand mortar (1:6 by volume) (Figure 4) showed satisfactory performance as a mortar. RHA cement was not recommended for reinforced concrete work.

Production Plant A1 (using technology of Institute A)

Approximately eight workers were seen to be busily engaged. Eight incinerators were being operated, in spite of rain and wind at the time of the visit, which made working conditions difficult (Figure 5). It was observed that unburnt husk was in and around the bottom area of some incinerators, yet fortunately the workers managed to avoid collecting unburnt material with the ash. It was said that a thermocouple enabled burning temperatures to be monitored and controlled, but there was some difficulty in obtaining a meaningful reading with the instrument when placed in a pile of hot ash lying in a storage shed.

Two brick-built lime kilns were at the site. One was to the design of the Khadi and Village Industries Commission, but was not used, since the workers preferred the other which was built to their own traditional design. It was 3.25m high, approximately 2 m diameter internally at the top, tapering to 1.5 m at the base. No lime burning was possible at the time of the visit because limestone deliveries from the quarries had been disrupted by landslides which had dislocated road transport. A simple flat area was available for slaking quicklime with water.

The ball mill (Figure 6) was apparently in working order, though unfortunately the electricity supply had been temporarily interrupted. Consequently no product was being made at the time of the visit, though a sample was obtained, and a representative sample of ash was taken from the storage shed.

When the product was being made it was sold and used locally, though some was transported to the nearest city. Its performance was said to be satisfactory and when inspected, it did appear so. It had been used for construction both in the plant and in the adjoining rice mill for mortars, renderings and for construction of a flight of steps.

An amorphous ash, obtained by burning husks at less than 750°C, was claimed to be desirable, but use of any type of ash which might become available was considered. This was because rice husk was being increasingly used as a fuel, so that it might not be cheaply available in a few years' time for burning in special incinerators. Consequently a technique which could be modified according to prevailing conditions was advocated. Although 2 hours grinding was adequate for amorphous ash, longer periods of grinding, of 6 hours or so, could be effective in increasing the strength of cements made from more crystalline ash. To conserve energy and improve efficiency of production, appropriate minimum grinding times should be used for each particular ash. Up to 10% carbon had not proved deleterious.

Where husk was available, controlled burning at a sufficiently low temperature would give a highly reactive ash which needed little grinding in order to obtain adequate cement strengths. A cylindrical metal wire mesh basket, 1.2 m high and 0.6 m in diameter, lined with a finer wire netting to retain husks, set in a mild steel frame on legs, and with a perforated metal chimney up the centre had proved successful as an incinerator (Figure 7). To operate, a fire was kindled at the base between the legs. A damper was provided at the chimney base so that with it open, air could flow in at the base or with it closed, air would be drawn in through the husks in the basket. The fire was allowed to burn up through the husks, volatile matter from them entering the chimney and shooting out of the top in a large flame. As husk burnt the top level fell, and was topped up with more husk. The whole lot was burnt out in less than two hours. In theory the volatiles could be taken off from the chimney for combustion elsewhere, though ducting would diminish the draught through the fire. However this might be the basis of utilising the fuel value of the husks. Alternatively, rows of such incinerators might be operated under some device which needed heating (e.g. in the drying of food crops). There was no convenient method of removing ash and the incinerators had to be tipped over by hand.

The BET method for determining specific surface area of husk burnt in a laboratory furnace had shown a decrease in area with increasing burning temperature. After determining the surface area of the ash, it was inferred from this relationship that temperatures had been approximately $400^{\circ}C$, but no direct temperature readings seem to have been taken from the incinerator.

Ash was used for two different products. One was produced by grinding ash and mixing with ordinary Portland cement (OPC). The other was made by grinding ash and hydrated lime together. OPC and lime were bought as available on the market.

The suggested size of ball mill for both products was 1.5 x 1.5 m, with a 25kW motor, capable of producing at least 1.7 tonnes per day, or more for a soft ash. A grinding medium of 3 tonnes of cast iron or steel balls, 20-25 mm in diameter was used, but no special mill lining was specified.

The lime-ash material was said to be the slower setting of the two products, but accelerators, other than up to 10% OPC, were not recommended as they would be an added complication and expense.

Proportions of lime and ash varied; for every part by weight of lime, from 1 to 2.5 of ash might be added depending upon requirements in use. It was satisfactory for masonry and foundations or general concreting work, but never for reinforced concrete. Proportions of OPC to ash could be one part by weight OPC with from 0.5 to 1.5 parts ash, depending whether high or ordinary strength was required. Instead of OPC, Portland pozzolana cement (PPC) may be used. PPC is OPC containing brick dust or other pozzolanic material, and one part would be mixed with from one to two parts ash by weight, depending upon whether strong or normal properties were needed.

For practical use as mortars, the lime-ash material would be mixed with sand in volume ratios from 1:1.5 to 4 and the OPC-ash material from 1:2 to 5 depending upon properties required.

It was emphasised that in practical use it should be remembered that although these cementing materials are sold by weight they are less dense than OPC. The lime-ash material in particular has a bulk density approximately half that of OPC. So there will be have to be nearly twice the volume to obtain the same weight ratio to sand and aggregate. Cubes had been cast and tested to ascertain the properties obtained when used in construction. Strength after 28 days was fairly high, a 1:3 mix of the OPC-ash material to sand being stronger than 1:3 lime-ash to sand. Pieces remaining from the test were inspected and appeared to be of good quality.

It cost approximately £30 to produce a tonne of either product which might be sold at £50. Actual and comparative costs with OPC were difficult to make at the time of the visit due to the recent change in marketing policy of the OPC as mentioned previously, but as OPC prices on the free market were rising at the time, RHA cements apparently gained an advantage.

The suggested production plant was similar in capacity to the plant A. The following costs presented here had not been updated for a few years so will be low. However, of the fixed capital required the ball mill cost at £4,500 and land at £1,200 were similar to the figures provided for Institute A. Allocation for buidings etc at £1200 was not directly comparable with A as boundary walls etc., were not mentioned. Total fixed capital was therefore approximately £7,000.

Estimates of annual costs of running the plant for 300 days per year, on three shifts per day for three workers per shift (less workers than A as ash is bought ready-burnt) plus materials, power and packing were difficult to assess with varying raw material prices, but may be between £8,000 and £14,000. Higher cement-content products will increase cost.

Indian patent no: 142966 existed on the process. Many enquiries about the process were reported and a number of entrepreneurs had commenced or were planning to commence manufacture in various places, within four separate States.

Of these only two were stated to be in production and there was some doubt about the others. Production of these plants was generally the lime-ash material.

In addition there were two other plants in operating condition and it was intended that they should both become not only production plants, but also demonstration and training centres for further developments.

Numerous enquiries had been received from India and from overseas. The following criteria were suggested before considering development.

1.   Is there the ability to absorb the technology?
2.   Does the necessary infrastructure exist?
3.   Does sufficient need for the product exist?

Production Plant B1 (using technology of Institute B)

The plant had three entrepreneurs who engaged eight workers.

At the time of the visit husks were being burnt (Figure 8) in a large open pile 2.5 m high, partly contained by a circular brick wall 16 m in diameter. The consultant from Institute B claimed to have advised against this procedure which he forecast would give ash with low reactivity. A quarter of the contents in the whole enclosed area was removed each day and replaced with new husk, though the extreme bottom of the pile was never cleared out. The fire was thus allowed to move round the area from one quarter to the next. It was said that wind direction had no effect.

Some ready-burnt boiler ash was sometimes brought in to supplement the supply, but was further burnt by laying it on top of the already burning husks in the incineration area.

The proprietors said they were planning to construct a large brick built type of incinerator.

The ash was normally ground by ball milling for five hours, then mixed with lime or cement and milled again, but prior to the visit the mill had broken down and had been removed elsewhere for repair. Two new smaller mills, each 1.2 x 1.2 m were seen being installed (Figure 9), but as they were not yet operational, no actual product was being made for sale at the time of the visit.

When in production 1 part of lime was mixed with 1.5 parts of ash by weight, and setting accelerated by the addition of secret additives. These were probably common salts, but no statement was made by the proprietors. OPC was used instead of lime, in the same proportion, for an alternative product. An 'Instructions for Use' leaflet in English and Hindi was included with every bag. Due to difficulties in obtaining OPC, some product was made using 1 part of PPC, with 1 part ash.

Current prices were said to be £40 and £52 per tonne for the two types respectively. No detailed information was obtained on economics of the plant, but it was said to be profitable, and the partners were planning to increase production.

Mix proportions for mortar and concrete were clearly stated - by weight 1:4 cement:sand and 1:2:4 cement:sand:coarse aggregate and by volume 1.75:4 and 1.75:2:4 respectively. 50% less mixing water than required for OPC was indicated.

The products from the works were said to be much in demand.

Research Institute C

In the process developed during the preceding 5 years, ready-burnt husk ash was bought usually from rice mills where it had been burnt in 1 metre high piles. Alternatively ash from husks burnt for parboiling of rice might be used. Two parts by weight of ash were mixed with one part lime and one part waste from a local industry. This waste gave higher strengths to the finished product, possibly due to the contribution of aluminates in the waste. Lime was bought in for pilot scale work at the Institute. The constituents were ground together for 3 hours in a 0.5 x 0.5 m rod mill, capable of giving 0.5 tonne per day.

Research into the strength of concrete specimens made with this product had led to the conclusion that $500^{\circ}C$ was a suitable ignition temperature for the husks though $700^{\circ}C$ was acceptable. Furthermore up to 20% of carbon in the ash seemed to be acceptable. Test results also indicated that concretes of 1:2:4 RHA cement:lime:sand, although weaker at 7 days, were stronger at 28 days than if OPC were used.

Some reinforced concrete samples had been kept in water for up to two years' soaking, but not enough was known at the time to make any recommendations.

The rice husk cement had been used successfully for making the rendering on the small building which housed the rod mill (Figure 10).

The cost of production on a commercial scale of 16 tonnes per day was said to be £28 per tonne, and the selling price was £36 per tonne. Eight people would be required and the cost of setting up such a plant, complete with lime kiln, was said to be £30,000 plus £12,000 for land and buildings.

Entrepreneurs were required to pay £300 for the 'know-how', then 1% of the sales for the first 5 years. For this payment entrepreneurs received a one-week training course, and a written document describing the process, detailing procedures for various chemical and other tests, and giving a sample economic feasibility report. The report indicated that a plant of 16 tonnes per day capacity, operating 24 hours a day, 300 days a year, would require 2 ball mills, costing a total of £18,000, other machinery and fittings £3,000, pre-operative expenses of £4,000, contingencies £4,000, working capital for salaries and materials for one month's production, stock, stores etc., £21,000. The cost of a lime kiln was not mentioned in these data.

Entrepreneurs who had gained knowledge and experience by working at the Institute were building production plants at two locations. At one, accommodation buildings were still under construction. However, adjacent to the Institute construction had been finished and equipment installed, and this latter plant was visited. Three more parties had the know-how and had been trained. They planned to set up plants so there would be production in three States.

Production Plant C1 (using technology of Institute C)

To enable ex-students from Institute C to obtain practical industrial experience in management and manufacturing, a division of the Institute has set up 30 small factory buildings, complete with services, on an adjacent site. One of the units is for the production of rice husk ash cement. Equipment and materials (Figure 11) had been assembled ready for commissioning a few days before the visit, but unfortunately the 1.8 x 1.8 ball mill was found to have misaligned bearings when it was first run, so it was being dismantled at the time of the visit. Such a mill was too heavy to man-handle, so shear legs with lifting tackle had to be brought into lift it. The mill was to take a charge of 2 tonnes, plus the weight of the balls. Husk ash was to be bought in ready-burnt, but two lime kilns had been constructed just outside the factory building. Other additives were seen piled against the factory wall. Proportions were to be measured out with gauge boxes, and a heavy steel truck was installed for removing the charge after milling. It was planned to automate parts of the process later. The whole process plant seemed well laid out, and the entrepreneur was enthusiastic, well-trained, and in a convenient position to call on the advice of the Institute's staff.

(This plant was reported to be working now.)

Research Institute D

The process utilised not only rice husk, but also lime sludge waste from the sugar industry. The lime sludge was a gel, which had to be dried first, mixed with an equal weight of husk, rewetted, then rolled by hand into 80 mm diameter balls. These were allowed to dry in the sun (Figure 12), then laid in a long trench. A fire was started at one end and allowed to burn right through to the other. The cooled balls were then ground for one hour to give a product which would be marketed as a lime-pozzolana.

Approximately $700^{\circ}C$ was reckoned to be a critical temperature for burning the husk. Below that temperature amorphous ash was produced, regardless of time of heating. Above, time of heating was important, so that more than 3 hours at $800^{\circ}C$ or more than 1.5 hours at $900^{\circ}C$ would give crystalline ash. However, one member of the staff reckoned the ash was 50% crystalline at $700^{\circ}C$ and $750^{\circ}C$ was the absolute maximum permissible. To convert the lime in the lime sludge into quicklime, a temperature of over $800^{\circ}C$ was necessary. The crude trench burning method was claimed to give sufficient temperature, though actual measurements showed only $700^{\circ}C$. The trench method allowed excess air, which helped to burn off carbon from the husk, and to remove carbon dioxide from the decomposing lime. 12% residual carbon gave maximum reactivity. Burning time is adjusted to give a reasonably white ash.

Setting times of 0.5 hour initial and 10 hours final were reported. Concrete made from the RHA cements (1:2:4 RHA cement:sand:coarse aggregate) was less strong than when ordinary Portland cement was used. Where no fired clay bricks were available, RHA cement was suggested as a basis for concrete block manufacture.

Estimates for production cost had been made prior to commercial application. Land, buildings, machinery etc., contributed to a capital cost of £10,000 and a similar amount might be required to meet annual costs.

The rendering (1:3 RHA cement:sand) on a brick wall, fully exposed to the weather for six years was inspected and found to be in good condition.

Some entrepreneurs were reported to have done experiments, and it was hoped that one would commence production within a few months but at the time of the visit there had been no commercial exploitation of the process.

2.2   NEPAL

Research Institute E

The method used involved collection of ash (from parboiling of rice etc.,) grinding in a ball mill, then further intergrinding for a total time of two hours with one third by weight of bought-in lime. Addition of 5-20% by weight of burnt clay had been found to give a more workable mix, but much grinding was required. A small proportion of ordinary Portland cement was added sometimes.

The product was not recommended for reinforced concrete construction.

Mortar could be made by mixing the product with three parts sand. However it was said that an equivalent low strength mortar could be made with OPC, using eight parts of sand, so to compete, the price of RHA cement must be much lower than that of OPC. In Kathmandu RHA cement sold at £55 per tonne while OPC was £90/tonne. It was claimed that Government contracts specified OPC and never RHA cement, so reducing the chances of the use of the latter. However, the current demand for cement in Nepal was said to be 300,000 tonnes per year; only 40,000 tonnes was produced locally. Demand should not fall since the Government's five year plan for 1985-90 was to concentrate on house building. Bringing in imported cement, over the difficult terrain resulted in high prices. These would be even higher in more remote parts, so there was considerable scope for small-scale RHA cement production to make up the deficit. A new

OPC plant of 120,000 tonnes per year was to be built in the south of the country, but even so there would be a shortfall in production, compared with demand.

The Institute had basic laboratory facilities and some RHA cement test results were reported, indicating reasonably good properties.

Twenty applications for details of the production process were said to have been received from entrepreneurs. Two were already said to be operating within Kathmandu, and three others were reported. The product was said to be selling well. The Institute reckoned it would cost £10,000 to set up a commercial plant.

Production Plant E1 (using technology of Institute E)

Rice grains for human consumption were being roasted on small husk-fire ovens. The husk ash was raked into piles out-of-doors (Figure 13), from where it was taken, while still hot, in wheelbarrows to the RHA cement production plant.

Ash was stored for two days and changed from dark grey to white. A small ball mill (0.7 x 0.7m) (Figure 14) was used to first grind the ash, then to intermix with lime, for a total time of two hours. The resulting fine powder was weighed into any available bags for sale (Figure 15) at £55 per tonne. Recommended mix for use was 1:3 RHA cement:sand.

Setting times of 1.5 hours initial and 3.5 hours final were claimed. A water requirement of 40% was quoted. Only about half a tonne could be produced per day, but it sold well.

Production Plant E2

An entrepreneur commenced production of RHA cement three years previously and had spent over £6,000 to date. He had bought 'know-how' of an early method that Institute E had apparently offered. The method consisted of making balls from rice husks and lime sludge, so was similar to that seen at Institute D, but the entrepreneur had no success with the method so changed to another. He built two large circular brick and concrete domed kilns, 3m high and 6m in diameter (Figure 16). The ash was mauve in colour, and the only way it could be removed was by someone entering the

-14-

dusty inside of the kilns. The ash was ground in
three locally-made ball mills of each 0.9m long
and 0.6m in diameter (Figure 17). Ball mill life
was said to be two years. Lime and ash were mixed
and put in gunny bags for sale. Maximum
production rate was 0.75 tonne per day. A stock
of some 9 tonnes was seen, which might deteriorate
with long periods of storage.

The RHA cement had been sold at £60 per tonne,
somewhat cheaper than ordinary Portland cement,
quoted as costing £90 per tonne.

For use, a 1:3 by volume RHA cement:sand mix was
usually suggested. The cement could be used for a
mortar, render and flooring, but not for
reinforced concrete work.

Since the properties were poor, and it was likely
that ash temperatures were too high, it was
suggested that husks be burnt in much smaller
quantities, in the first instance by dividing an
existing kiln into several compartments.
Temperatures might be monitored with a
thermocouple.

2.3   PAKISTAN

Research Institute F

Rice husk was being used increasingly as a fuel,
for example half of the existing brick kilns were
said to use it, so it was becoming less abundant
and more expensive. Three new OPC plants were
planned so that eventually there would be a
surplus for export.

Nonetheless, there was a case for production of
RHA cement. Firstly it could have specific
properties, such as better chemical resistance
than OPC, so it could be of particular use, for
example for floors in the food industry where acid
resistance would be a benefit. It was said that
in big towns businessmen were not interested in
small investments, but for special-purpose cement
making, larger plants might be considered.
Secondly, in remote rural areas there was a need
for cementitious materials, but there were
increased transport difficulties and costs.
However, rural people had a fondness for locally
made things, so small scale RHA cement may be very
acceptable in that situation. Research had been
undertaken at laboratories and put into full scale
practice by an Agricultural Corporation (see

below), but it was hoped that the method could be tried out in a model village to build confidence with the people, then presented to others as a proven package. The estimated cost of setting up was only £1,600, and might be taken up by cooperatives in the rural areas. It was thought that RHA cement would be made for £10 per tonne, which was much less than the current selling price of OPC at £100 per tonne, giving a good margin for error and profit.

Husks were burnt in special incinerators made from oil drums, or in large heaps, but in either case the temperatures were said to be limited to 600 to 700°C. After grinding, the ash was mixed with lime.

The Institute's extensive laboratory facilities were visited. Fast burning was found to give black ash, which became white on standing. It was claimed that more than 4% carbon in the ash resulted in a loss of strength in use, though workers elsewhere have found that even 15% is acceptable. It had been found that an addition of 10% of OPC gave better workability.

Adjacent to the laboratories were well-equipped workshops. Two ball mills were in the course of manufacture, each 1.3 x 1.3m, which might be used for work on RHA cement (Figure 18).

Production Plant F1 (using technology of Institute F)

The Agriculture Corporation had been producing a large quantity of rice husk and sought a useful method of disposal. Consequently they set up a plant which was said to cost £1,600 and could produce 3 tonnes per day of rice husk ash cement. This was for use in their own construction programme.

The husk had been burnt in a heap 2.5m high, at 600°C (checked with a thermocouple). Ash was ground in two ball mills each 0.9 x 0.8m for 5 to 6 hours, using hard steel balls as the grinding medium, then 30% by weight of lime was added. The resulting RHA cement was used in the proportions of 1:3 RHA cement:sand.

However, the demand for husks as fuel increased and it was found to be economic to sell the husks rather than make RHA cement, so the plant ceased operation after only one year, during which time only 100 tonnes was produced. Consequently the plant was not visited.

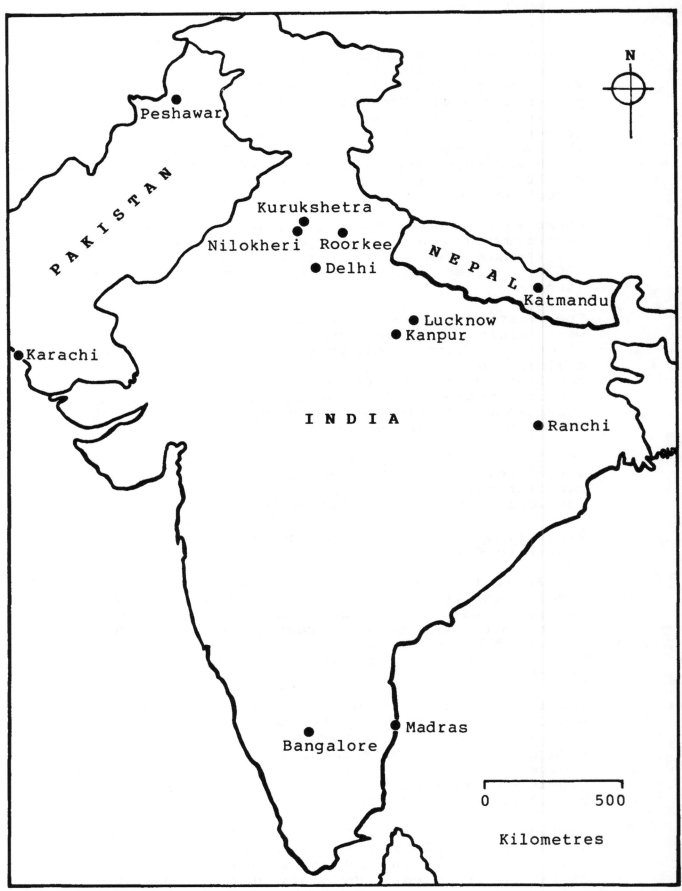

Figure 1. Some Centres of Rice Husk Ash Cement Information,
Research and Development in the Indian Sub–Continent

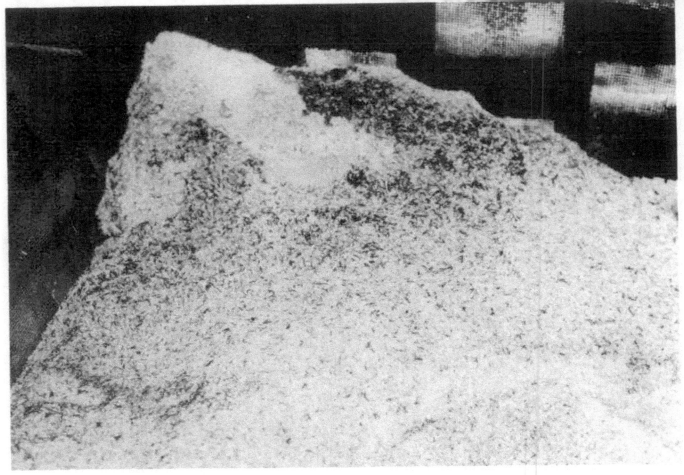

Figure 3. Burnt husk, retained by mosquito netting in open brickwork incinerators at Ballabgah

Figure 4. Fair-faced brickwork using rice husk ash cement mortar, at Ballabgah

Figure 5. Rice husk incinerators at Nilokheri

Figure 6. Ball mill at Nilokheri

Figure 7. Prof. P.C. Kapur with TiB incinerator at Kanpur

Figure 8. Heap burning of rice husk at Kurukshetra

Figure 9. Mr. Lalit with new ball mill at Kurukshetra

Figure 10. Rice husk ash cement rendering at Institute C

Figure 11. Nearly-complete production unit at Plant C1.

Figure 12. Dr. A. Dass at CBRI shows lime sludge/rice husk balls laid out to dry

Figure 13. Rice husk ash is raked out from small oven in Kathmandu

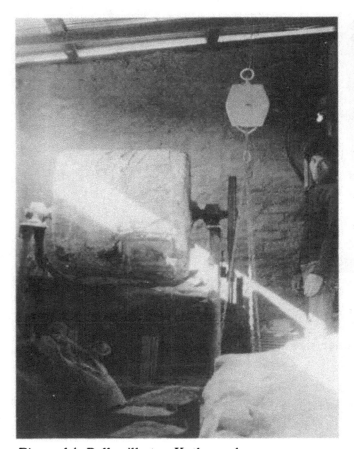
Figure 14. Ball mill etc., Kathmandu

Figure 15. Author and Mr. R.D. Shrestha with bagged cement for sale at street-side

Figure 16. Large incinerator at Patan

Figure 17. Mr. M.P. Dabadi with three ball mills at Patan

3.   COMPARISON OF PRODUCTION METHODS AND
      RECOMMENDATIONS FOR USE

To facilitate a simple comparison of the methods,
Table 1 shows the size of plants, usual sources of
the basic raw materials, grinding time in the ball
mill, and number of employees. Data can only be
regarded as approximate in most cases. The wide
variation in grinding time at Institute B is by
design: soft ash required shorter time than hard
ash. Many plants were not completely operational
at the time of the visit, though methods were
discussed and some equipment and materials
inspected, as detailed above.

Plant capacity was very small in all cases,
averaging approximately 4 tonnes per day.

Methods were nearly equally divided between those
burning husk in special incinerators and those
buying in ready-burnt material, but most operators
bought ready-burnt lime. OPC and aluminates were
always bought ready made.

Grinding time varied by several hours about a mean
of 3.5 hours.

The numbers of employees was difficult to assess
in some cases, but averaged approximately seven
per plant.

The success of the operations seemed to vary
immensely. Some were well run profitable
businesses, while others were losing money.
Research carried out in some institutes had been
applied by entrepreneurs, in other cases
entrepreneurs had installed equipment and were
ready to commence production. Some research had
not yet got beyond a basic agreement with an
entrepreneur, as detailed above.

Mix proportions used in manufacturing RHA cements
and for their use in mortars are summarised in
Table 2. In general one part of lime has been
mixed with two of RHA by weight, but with OPC and
aluminates one part has been mixed with only 1.5
of RHA. For use as mortar the recommended mix is
in the proportions of one part RHA cement to three
parts of sand by volume generally, but this varies
depending upon RHA cement quality and mortar
quality required.

4.   TESTING SAMPLES IN UK

Samples of ashes and RHA cements were collected where possible, and brought back to England for test under standard conditions. Ashes were examined by X-ray diffraction techniques to determine whether they were amorphous or crystalline. Ashes were mixed with half their weight of fresh hydrated lime and then tested in the same way as the ready-made RHA cements. Methods were based on those specified in the British Standard Specification for Cement, BS 12:1978. Soundness of the materials was done by the Le Chatelier method, initial and final setting times with the Vicat apparatus, using pastes, when sufficient material was available.

For compressive strength tests, cements were mixed with three times their volume of graded sand (BS 4451) and cast into 25mm cubes (smaller than BS 12). The quantity of water in each mix was that necessary to give a standard workability. In addition to testing after 7 and 28 days storage at ambient temperature (20°C), an accelerated test was carried out using an elevated temperature (50°C).

Table 3 identifies the nature of the samples taken at the various places visited. It is not possible to say how representative the samples were of production over a long period.

5.   RESULTS OF UK LABORATORY TESTS

Results of tests are divided into three separate groups. First the samples of 'as-sold' RHA cement made using local lime and ash, then those 'as-sold' using OPC etc., and ash, and finally the mixture of a standard UK lime with the ash. Results are given in Table 4.

5.1   Compressive strength

In accordance with common practice, 1:3 cement:sand mortar was made but because of small quantities of material available 25mm cubes were cast and cured at 20°C then tested at 7 and 28 days. In addition an accelerated test was carried out by keeping some cubes at an elevated temperature of 50°C from day 4 to day 7, cooling for one hour, then testing. At least three cubes were used for each test, and results showed very good reproducibility.

Of the 'as-sold' RHA cements, the lime/ash were not as strong as the OPC/ash in 7, 28 and 7 (accelerated) day tests. The apparent exception, lime/ash B probably owes its higher strength to the incorporation of a small proportion of OPC. The lime/ash 'as-sold' cements were not as strong as the mixtures of ashes with UK lime, in general, perhaps due to the relative freshness of the UK lime.

Inspection of data in Table 4 shows that 7-day accelerated test results give an approximate indication of 28-day strength under ambient test conditions.

Overall, the materials gain approximately three times their 7 day strength by 28 days. This ratio does not differ significantly between lime and aluminates, though where OPC is used in the mix there is slightly faster development of early strength.

5.2   Soundness

Le Chatelier's method, with a split ring mould and pointed indicators to amplify any movement, was used. No significant expansions were found in any of the samples tested.

5.3   Setting Time

The Vicat apparatus, with its penetrating needle and attachments was used. Initial setting times varied widely from 3 to 30 hours, and final setting times from 10 to 36 hours.

Within each of the two groups of 'as-sold' cements (i.e. the lime/ash and the OPC or aluminates/ash) the higher the compressive strength the faster were the initial and final sets.

5.4   Crystallinity

X-ray diffraction analysis of each ash sample showed that three were mainly amorphous while the others were crystalline. Results are detailed in Table 5.

6. COMPARISON OF PRODUCERS' DATA WITH UK RESULTS

The properties of RHA cements which were claimed for the various 'as-sold' products during the visit are listed in Table 6, which is in the same format as Table 4 for ease of comparison.

Compressive strengths claimed were up to about three times greater than those found in the UK tests in spite of the fact that smaller cubes used in the UK would give greater strength. An exception was the Institute C samples, at 28 days. Possibly the claimed higher values were partly due to materials being more fresh.

Soundness data reported by producers showed slightly greater expansions.

Setting times claimed were faster than UK results show. Again this could be due to material being more fresh. The E2 sample was an exception.

Some data was collected on fineness.

7. COMPARISON OF EXPERIMENTAL DATA WITH STANDARD REQUIREMENTS

There is no existing standard specification entirely relevant for rice husk ash cements. Table 7 shows the requirements of standards for materials intended for a similar use. The most relevant is IS 4098-1967, the Indian Standard for Lime-Pozzolana Mixture.

Producers' data for 'as-sold' lime/ash products meet the highest requirements of IS 4098 (LP 40), but UK tests on the same materials show some of them only satisfy the minimum requirement (LP 7). UK lime/ash mixes meet the highest requirement. Mixes containing OPC, being stronger, exceed the highest requirements of IS 4098, and some even exceed requirements for OPC. It is difficult to be certain of the exact composition of the samples, and as noted above in the accounts of site visits, it is known that aluminates, OPC and PPC are always used in some products, but they may be incorporated sometimes in others.

Initial setting times claimed by producers are near the requirement of IS 4098. UK tests show longer initial setting times, but this is allowable in the standard.

Final setting times claimed are quite fast, within the fastest limit of IS 4098. All UK tests are within the slowest rate of that standard.

Fineness of the RHA cements quoted by some manufacturers is finer than required for OPCs. IS 4098 requires 10% (maximum) retained on a 150-micron sieve, so available data cannot be compared directly with this.

8. ECONOMICS

The economics of the various production methods were difficult to compare, due to the different amounts of information, and the possible inaccuracy of information obtained. For example, the Institute A provided detailed costings, which applied to the commercial production unit A1, and although Institute B gave figures, the plant B1 was so different from the general model that the figures might be considered inapplicable and no data other than selling price was revealed by the entrepreneurs at B1. Costs of raw materials vary continually, and the numbers of workers is difficult to define accurately.

The price of OPC in India had been Government controlled, and kept uniform throughout the country until one month before the visit. A third of production was then released on to the free market, and its price escalated rapidly from £50 per tonne (controlled) to approximately £95 per tonne (free). This would appear to increase the selling price advantage of RHA cement, though it would increase production cost of those RHA cements containing OPC. Thus prices were in a state of flux at the time of the visit to India.

Nevertheless some comparative data is presented in Table 8. The cost of ash is generally low. In some cases it is only the cost of transportation of husk from the rice mill to incinerator. Some expenditure is inevitable in the burning process, but is not necessarily included in the figures obtained and quoted in the table.

Cost of lime must be important but would vary. For example at Plant E2 in Nepal the entrepreneur spoke of preferring Indian lime which was more expensive than local lime, so the higher figure is quoted on the table.

The cost of making RHA cement could not always be
obtained.   Where it is high, selling price is
high.

Selling price of RHA cement varied from £27 to
£60, with the exception of the Institute D price
which was hypothetical, as it was not actually in
commercial production.   The plant in Pakistan only
made   RHA   cement   for   use   by   the   owning
organisation, not for sale, so no selling price
could be quoted.

Prices   for   OPC   were   obtained   in   all   three
countries, though not at every location visited.
Overall, the selling price of RHA cement was
approximately 45% of the price of OPC.   However
there are two conflicting factors which should be
considered   as   they   affect   the   extent   of   this
apparent price difference in its effect on the
user of the product.   Firstly the bulk density of
RHA cement may be only two thirds that of OPC, so
for a mix specified by volume (as is common) less
weight of RHA cement will be required than for
OPC.   The cost advantage of RHA cement would be
improved therefore to 2/3 of 45% or 30% of OPC
cost.   Secondly, on the other hand, the strength
of   the   normally   suggested   1:3   mix   of   RHA
cement:sand for mortars (see above) could be
equalled by a less rich (say 1:7) OPC:sand mix.
Thus 2.3 times as much RHA cement may be required
for an equivalent mix, bringing the cost advantage
to the user of RHA cement to 2.3 times 30% or 70%
of OPC cost.

Capital cost for setting up plant was obtained
from most places visited.   There was no way of
checking the details in some instances.   The
relatively low cost quoted for the Pakistan plant
is surprising, but may exclude some items of cost
included in the other figures.   Cost in Nepal at
E2   includes   cost   of   abortive   experimental
production.   Since plants are of different size
outputs, direct comparison of capital cost has
little meaning, so these figures may be divided by
the plant capacity, giving for India a capital
cost per tonne per day of between £3,000 and
£4,000.

Net annual cost figures were obtained from some
places and found to be of the same order as the
capital cost.   For comparison these net annual
cost figures may be divided by the annual tonnage
produced, when it is seen that net annual cost is

approximately £20 per tonne; individual figures agree with the cost of making one tonne quoted in the same table.

More detailed calculations by two of the Institutes lead to calculated returns on investments of 32% and 20%.

9.   CONCLUSIONS

1.   It has been amply demonstrated that RHA cement can be made using simple technology not only on laboratory or pilot scale, but also as a commercial venture producing several tonnes of product per day.

2.   To maintain production, supplies of raw materials must be continuously available or stock-piled. Power must be available especially for ball mills, with no more than infrequent short interruptions, or ball mills must be over-size capacity to allow for these interruptions.

3.   To maintain sales of the product, especially when alternative materials are available, the RHA cement must be of a consistently high quality. It is best to burn husk at no more than 700°C.

4.   Several variations of the basic process are available. Ash and lime alone, which constitute a type of lime-pozzolana mixture, can give a useful cementing material. When particular local conditions permit, the basic mixture of ash and lime may be modified with advantage.

5.   Controlled burning of husk in specially made incinerators though desirable, is not essential to producing a good RHA cement. However it is probably desirable in order to get a cement of consistent properties. Chemical analysis and fineness determinations might have been done on these samples with advantage in helping to understand their behaviour.

6.   Husk burnt in large quantities is likely to be crystalline. Amorphous ash is produced when smaller quantities are burnt with plenty of air. 10 to 20% of carbon in the ash may be acceptable.

-34-

7.   If ash of various qualities is to be accepted, it may be necessary to adjust grinding times in order to develop sufficient strength in mixes with time.

8.   The ash from plant A1, though amorphous and said to be ground very fine, produced surprisingly low strength mortars. The crystalline ash from B1 produced surprisingly high strength mortars.

9.   The RHA cement has properties which make it suitable for mortars and rendering, that is to say for use as a masonry cement. Some of the products tested meet performance requirements of BS 5224:1976. Some meet IS4098:1967. Concrete for general work could be made, though strength may not be as high as with OPC. There is no evidence that RHA cement concrete allows corrosion of embedded steel but at the present time it is not recommended for reinforced concrete work.

10.   RHA cement may cost only 45% of the cost of OPC in the market. Mortar made with RHA cement may cost only 70% of equivalent OPC mortars.

11.   There may be additional technical advantages in some possible areas of use of RHA cements. Heat generated during hydration may be less than that of OPC, so lower thermal stresses would result if RHA cements were used in massive concrete structures. Where concrete aggregates may be reactive with lime released during setting of OPC, which in some instances has lead to disruptive stresses being set up, RHA cements may possibly show less reaction with those aggregates. Chemical resistance, to acids and sulphates, may be enhanced if RHA cements are used instead of OPC.

12.   Guidelines for design of 4-tonne per day
        small-scale plant burning its own limestone
        and husk to make RHA cement:

Capital investment	£12,000
Net annual cost	£12,000
Workforce	8 people
Materials	12 tonnes rice husk per day
	2 tonnes limestone per day
Equipment to include	small incinerators with open-work sides, lime kiln ball-mill, operating for 3 hours per batch
Selling price	£40 per tonne
Mix proportions	RHA cement may consist of 1 part lime - 2 parts ash by weight, mortars may be made from 1 part RHA cement - 3 parts sand by volume.

Detail will vary considerably, depending upon
local circumstances.

10.   RECOMMENDATIONS

1.   Incinerators for burning husk should allow
        plenty of cooling air which will also help
        to keep carbon content low.   This is more
        easily achieved in small incinerators than
        in large heap burning.   The temperature
        should be limited to $700^{\circ}C$ maximum to avoid
        formation of crystalline ash.

2.   Heat generated by combustion of ash should
        be usefully employed, for example for
        drying foodstuffs.

3.   The period of grinding in the ball mill
        should be adjusted to the minimum necessary
        to produce the required properties in the
        RHA cement.   Specially burnt ash should be
        soft and require less grinding, but if ash
        is bought in ready burnt, having already
        contributed its heating value in some
        process, longer grinding may be necessary
        in order to produce a satisfatory RHA
        cement.

4.   Inhalation of dust should be avoided as far
        as possible, since the extent of the health
        risk is not fully understood at present.

5.   At present it would be advisable to market
        the material as a cement for masonry.
        Packaging should be clearly marked to
        differentiate RHA cement from other types
        of cement.

6.   Liaison should continue with the institutes
        and entrepreneurs already visited.
        Institutes will be in possession of new
        information. Some entrepreneurs had only
        just commenced production, others had not
        started, and one was recommended to make
        changes, at the time of the visit, so
        subsequent production results would be of
        interest. Further advice might be welcomed
        by entrepreneurs.

7.   Information should be sought on the
        unformity of product quality from various
        plants over a long period. Quality should
        be recorded and producing countries should
        compile standard specifications.

8.   New areas for application of this
        technology should be found in other rice
        growing countries where there is a shortage
        of cementitious materials.

TABLE 1   COMPARISON OF PRODUCTION METHODS INVESTIGATED

Country	Location	Scale of production	Plant capacity (tonnes) per day	Raw Materials and sources — Ash	Raw Materials and sources — Lime	Raw Materials and sources — OPC etc	Grinding Period in ball mill (hours) — Ash only	Grinding Period in ball mill (hours) — Total	Number of employees	Notes (at time of visit)
India	A1	Commercial	2	S	S	–	1.5	2	8	Production stopped temporarily
	B	Pilot	1.5	S	B	B	–	2 – 6	–	Not being used
	B1	Commercial	8	S*	B	B	5	5.5	8	Production delayed temporarily
	C	Pilot	0.5	B	B	B	–	3	–	Not running
	C1	Commercial	16	B	S	B	–	3	8	Not yet in production
	D	Pilot	5	S	B	–	–	1	15	No longer in production
Nepal	E1	Very Small Commercial	0.5	b	B (dried sludge)	–	1.5	2–3.5	2	Working
	E2	Commercial	0.75	B	B	–	–	probably 6	3	Working but product not selling
Pakistan	F1	Commercial	3	B	B	–	–	5 – 6	–	No longer in production

B = Bought-in ready-burnt from large scale burning elsewhere.
b = Bought-in ready-burnt from small scale burning elsewhere.
S = Specially burnt in purpose-built incinerator or kiln at RHA cement plant.

* Some B

TABLE 2   MIX PROPORTIONS USED IN MANUFACTURING RHA CEMENT AND IN USE AS MORTARS

| Country | Location | Proportions of constituents in RHA cement (by weight) | | Proportions of mortar (by volume) |
		Lime:ash	OPC:ash	rha cement: sand
India	A1	1 : 2		1 : 4
	B	1 : 2.5		1 : 1.5 to 4
	B1	1 : 1.5	1 : 0.5 to 1.5	1 : 2 to 5
		plus additives		
	C		1 : 1.5	1 : 2.25
	D	1(lime sludge) : 1	1*: 1	1 : 3.5 to 4.5
				1 : 3
Nepal	E1	1 : 3		1 : 3
	E2			1 : 3
Pakistan	F1	1 : 2.25		1 : 3

* Not OPC, but equal weights of lime and an aluminate waste.

TABLE 3 RICE HUSK ASH MATERIALS SAMPLED

Country	Location	Ash	Lime/Ash cement	OPC or aluminate/ Ash cement
India	A1	2	1	–
	B	12	11	10
	B1	13	–	3
	C	–	–	15,16
Nepal	E1	7	6	–
	E2	8	4(with brick dust)	–
			9	–

Type of sample with sample reference number

TABLE 4   PROPERTIES OF RHA CEMENTS, DETERMINED IN UK LABORATORIES

Type and source of sample	Sample No.	Compressive Strength (MN/m²) 25 mm mortar cubes			Soundness Le Chatelier Expansion (mm)	Setting time (hours) Vicat	
		7 day 20°C	28 day 20°C	7 day 50°C		Initial	Final
1) Lime/ash products, as sold							
India A1 (a)	1	0.5	2.2	1.6	-	30	40
B	11	5.7	14.5	14.7	-	-	-
Nepal E1	6	1.5	6.2	3.2	0.5	12	20
E1 (20% brick dust)	4	0.9	4.3	3.5	1	10	24
E2	9	0.34	3.8	1.7	1	10	24
2) OPC or aluminates/ash products as sold							
India B	10	10.0	16.6	16.5	-	-	-
B1 (with PPC)	3	10.3	15.6	15.2	-	3	10
C – slow set	15	3.6	10.7	8.9	0.5	28	36
C – normal set	16	4.4	14.5	10.4	0.5	3	24
3) UK lime/ash							
India A1	2	2.1	4.6	3.9	-	-	-
B (specially burnt ash)	12	8.0	10.5	9.1	-	-	-
B1	13	3.4	7.2	6.2	1	16	24
Nepal E1	7	6.4	10.6	9.5	-	-	-
E2	8	0.07(b)	0.41	0.96	-	-	-

(a) may include 10% OPC
(b) not all cubes tested successfully

TABLE 5 RESULTS OF X-RAY DIFFRACTION ANALYSIS OF RICE HUSK ASHES

Country	Location	Sample No.	State and Minerals identified in Ash
INDIA	A1	2	Mainly AMORPHOUS plus minor crystalline quartz
	B	12	Mainly AMORPHOUS plus minor crystalline quartz
	B1	13	CRYSTALLINE
NEPAL	E1	7	Mainly AMORPHOUS plus minor crystalline quartz
	E2	8	Mainly CRYSTALLINE tridymite (SiO_2 polymorph)

TABLE 6   PROPERTIES CLAIMED BY PRODUCERS

Type of Cement and Producer	Compressive Strength (MN/m²) 50 mm mortar cubes		Soundness Le Chatelier Expansion (mm)	Setting Time (Hours) Vicat		Fineness Blaine (m²/kg)
	7 day	28 day		Initial	Final	
1. Lime/RHA cement product						
INDIA A1(1)	-	4	-	2 to 4	-	1000
B	15	25	-	1.5	6	-
D	6	11	2.5	0.5 to 1.5	8 to 10	500
NEPAL E1	6	15(2)	1	1.5	3.5	-
E2	-	8	-	-	48	-
PAKISTAN F1	16	19	0	3	4	400
2. OPC or Aluminate/RHA cement product						
INDIA B (PPC)	27	42	-	0.5	3.5	-
C	4	6	3	21	23	-
C (different composition)	6	10	1.5	2	3	-

1. May contain 10% OPC
2. 3 weeks, 1:2:4 concrete mix.

TABLE 7   REQUIREMENTS OF STANDARD SPECIFICATIONS

INDIAN STANDARDS (IS) AND BRITISH STANDARD (BS)

Standard	Compressive strength Mortar cubes (MN/m² minimum)			Soundness Le Chatelier Expansion (mm maximum)	Setting Time Vicat		Fineness Blaine (m²/kg minimum)
	7 day	14 day	28 day		Initial (hrs minimum)	Final (hrs maximum)	
IS 4098-1967 Type LP 40	2	-	4	Firm no cracking	2	24	-
Lime-pozzolana mixture Type LP 20	1	-	2		2	36	-
Type LP 7	0.3	-	0.7		2	48	-
IS 1489-1967 Portland-Pozzolana Cement	17.5	25	-	10	0.5	10	300
IS 269-1967 Ordinary Portland Cement	22	-	-	10	0.5	10	225
BS 12:1971	23	-	-	-	-	-	-
:1978 Ordinary Portland Cement	-	-	41	10	0.75	10	225

TABLE 8   ECONOMICS - COMPARISON OF VARIOUS PRODUCTION METHODS

Country	Location	Cost of ash £ per tonne	Cost of lime £ per tonne	Cost of making RHA cement £ per tonne	Selling price of RHA cement £ per tonne	Local Selling price of OPC £ per tonne	Capital Cost £	Capital Cost per tonne per day £	Net annual cost £	Net annual cost per tonne £	Return on Investment
India	A1	1	8	19	27	95	12,000	6,000	12,000	18	32%
	B			30	50	higher than	7,000	5,000	8,000 to 14,000	15 to 30	
	B1				40 to 52	the cement					
	C	3		28	36		42,000	2,500			
	D			10	13		10,000	2,000	10,000	10	20%
Nepal	E1				55	90	10,000				
	E2	10	40		60	90	6,000	8,000			
Pakistan	F1	2	15	10	not sold	100	1,600	500			

-45-